FITCH STREET SCHOOL LIBRARY

PROJECTS WITH
MACHINES

By
John Williams

Illustrated by
Malcolm S. Walker

Gareth Stevens Children's Books
MILWAUKEE

For a free color catalog describing Gareth Stevens' list of high-quality children's books, call 1-800-341-3569 (USA) or 1-800-461-9120 (Canada).

Titles in the Simple Science Projects series:

Simple Science Projects with Air
Simple Science Projects with Color and Light
Simple Science Projects with Electricity
Simple Science Projects with Flight
Simple Science Projects with Machines
Simple Science Projects with Time
Simple Science Projects with Water
Simple Science Projects with Wheels

Library of Congress Cataloging-in-Publication Data

Williams, John.
 Projects with machines / John Williams ; illustrated by Malcolm S. Walker.
      p. cm. -- (Simple science projects)
   Rev. ed. of: Machines. 1990.
   Includes bibliographical references and index.
   Summary: Provides instructions for making and using a variety of machines, including levers, diggers, cranes, and pulleys.
   ISBN 0-8368-0769-3
   1. Machinery--Experiments--Juvenile literature. [1. Machinery--Experiments. 2. Experiments.] I. Walker, Malcolm S., ill. II. Williams, John. Machines. III. Title. IV. Series: Williams, John. Simple science projects.
TJ147.W52 1992
621.8'078--dc20   91-50547

North American edition first published in 1992 by

Gareth Stevens Children's Books
1555 North RiverCenter Drive, Suite 201
Milwaukee, Wisconsin   53212, USA

U.S. edition copyright © 1992 by Gareth Stevens, Inc. First published as *Starting Technology — Machines* in Great Britain, copyright © 1991 by Wayland (Publishers) Limited. Additional end matter copyright © 1992 by Gareth Stevens, Inc.

Editor (U.K.): Anna Girling
Editors (U.S.): Eileen Foran, John D. Rateliff
Designer: Kudos Design Services
Cover design: Sharone Burris

Printed in Italy
Bound in the Unted States of America

1 2 3 4 5 6 7 8 9 97 96 95 94 93 92

CONTENTS

Simple Machines 4
Making Signals 6
Arms and Legs 8
Extenders 10
Catapults 12
Diggers 14
Pulleys 16
Cranes 18
Balloon Power 20
Dragons 22
Switches 24
A Switch Machine 26

What You'll Need 28
More Books About Machines 29
More Books With Projects 29
Places to Write for Science
   Supply Catalogs 29
Glossary 30
Index 32

Words printed in **boldface** type appear in the glossary on pages 30-31.

SIMPLE MACHINES

Machines help us do work. They may be very simple, like can openers, or they may be very complicated and large, like machines in **factories**.

All machines are based on simple ideas. **Levers** and **pulleys** are both simple machines. However big or complicated some machines may seem, they are really just a lot of simple ideas linked together.

In this factory the big, complicated machines are controlled by computers. They do not need many people to look after them.

Making a moving cat

You will need:

Cardboard
A pencil
Paper fasteners
Scissors
A paintbox
A ruler

1. Draw the shape of a cat's head and body on cardboard. Cut it out.

2. Draw two legs and a tail. Make sure the legs are at least 3/4 inch (2 cm) wide. Make the tail twice as long as the legs.

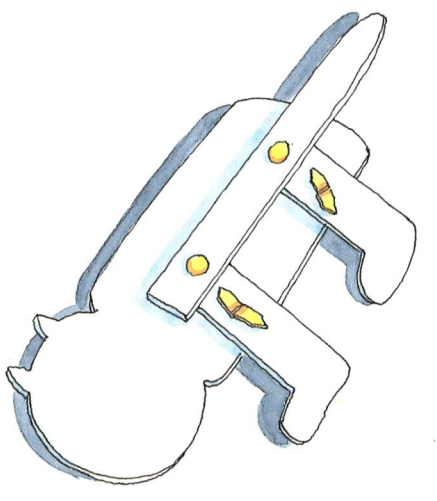

3. Use paper fasteners to attach the legs to the back of the cat's body. Attach the tail to the top of the legs, as shown. When you move the tail now, the legs will also move.

4. Paint the cat's face and body.

5. Make a cat that sits up like this one. Attach the front legs to the body, then connect the tail to the legs, as shown.

5

MAKING SIGNALS

Today, railroads use signal lights and complex computers to make sure all the trains are in the right place at the right time. In the past, railroads used signals like the one shown below. A system of levers moved the signals up and down. Some model train sets still use old-fashioned signals.

Old-fashioned railroad signals like this have been replaced by colored lights and computerized signals.

Making an old-fashioned railroad signal

You will need:

White cardboard
A pencil
A ruler
Four paper fasteners
Scissors
Red paint

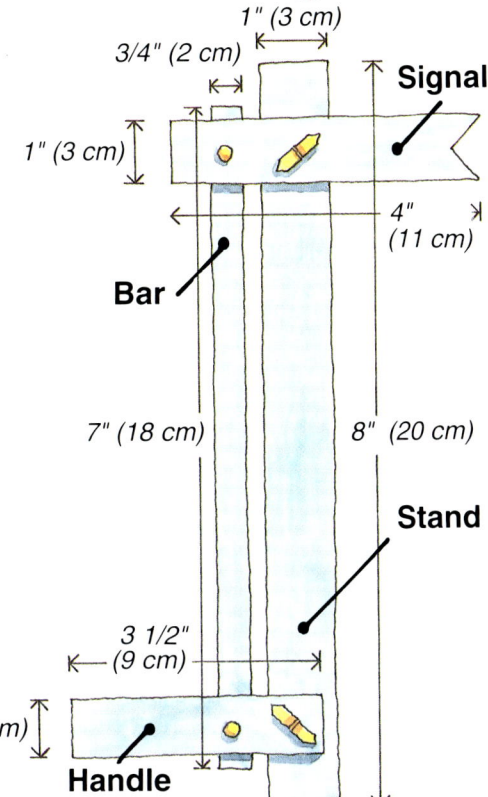

1. Cut four strips of cardboard. The stand should be 8 inches (20 cm) by 1 inch (3 cm). Make the bar 7 inches (18 cm) by 3/4 inch (2 cm). Make the signal 4 inches (11 cm) by 1 inch (3 cm). Make the handle 3 1/2 inches (9 cm) by 1 inch (3 cm).

2. Use paper fasteners to attach the signal and handle to each end of the bar. Next, attach them both to the stand, as shown. When you move the handle, the signal goes up and down.

3. Cut a V-shape in the end of the signal and paint a red stripe on it to make it look like an old-fashioned signal.

7

ARMS AND LEGS

Our bodies work like machines. Bones in our arms and legs work like levers. Our **joints** act as **fulcrums**. When you nod your head or open and shut your mouth, you are working a lever.

This funny robot was used as a character in a movie. It has mechanical arms that work like human arms. Can you see the joints and levers?

Making a model arm

You will need:

Cardboard
A pencil
A ruler
Paper fasteners
Tape
A rubber band
Scissors

1. Cut out two strips of cardboard 4 inches (10 cm) long and 3/4 inch (2 cm) wide. Cut out the shape of a hand and tape it to the end of one of the strips.

2. Use a paper fastener to join the two strips at the "elbow." Use the same fastener to attach the arm to a cardboard base. Now attach the "shoulder" to the base.

3. Tape the rubber band to the center of both parts of the arm. Try moving the lower part of the arm up and down. What happens?

4. Now make a person that has a cardboard body and rubber band muscles, like this.

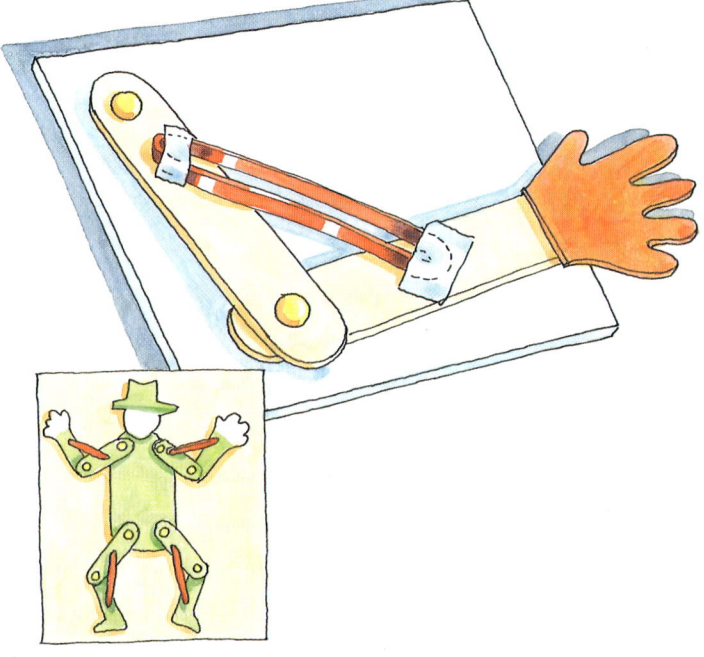

9

EXTENDERS

Making a mechanical extender

You will need:

Cardboard   Glue
Paper fasteners A ruler
Scissors   A pencil

1. Cut out eight strips of cardboard 6 inches (15 cm) long and 3/4 inch (2 cm) wide.

2. Push the fasteners through the strips to join them in the middle and at each end.

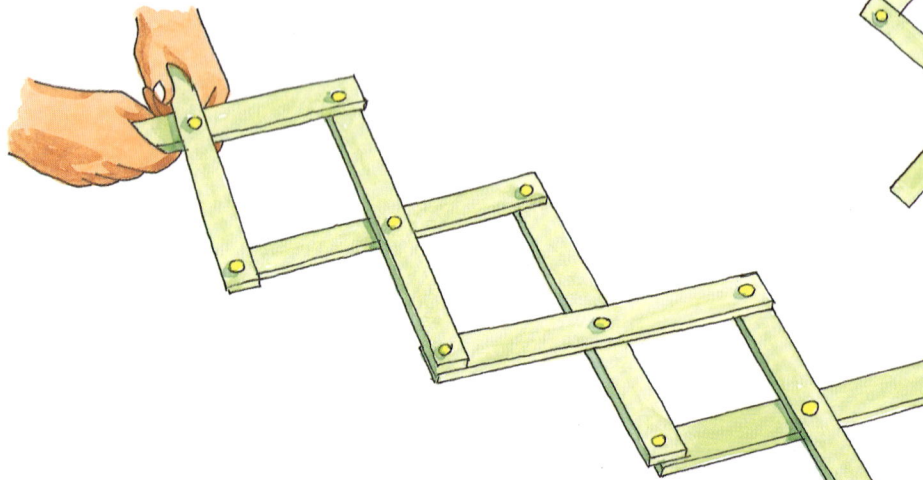

3. Glue two triangular-shaped "fingers" at one end of the arm. Now you can use your extender to pick up light objects.

Further work

Next time you are near a building site, look closely at the machines there. Do you see any machines like the mechanical extender you made on page 10?

This machine, known as a cherry picker, is made from many simple levers, one on top of another.

CATAPULTS

Catapults are machines used to make objects fly through the air. Long ago, armies used giant catapults to hurl stones at their enemies.

Making a catapult

You will need:

A block of wood
Two nails
A small hammer
A rubber band
A plastic spoon
A clothespin

A strip of cardboard
A thumbtack
Glue
Scissors
Modeling clay
A ruler

WARNING: Ask an adult to help you with the hammer and nails.

1. Hammer the nails firmly into the wooden block 4 inches (10 cm) apart. Leave at least 3/4 inch (2 cm) of each nail showing above the wood, as shown.

2. Stretch a twisted rubber band between the nails. Push the handle of the spoon through the twisted rubber band.

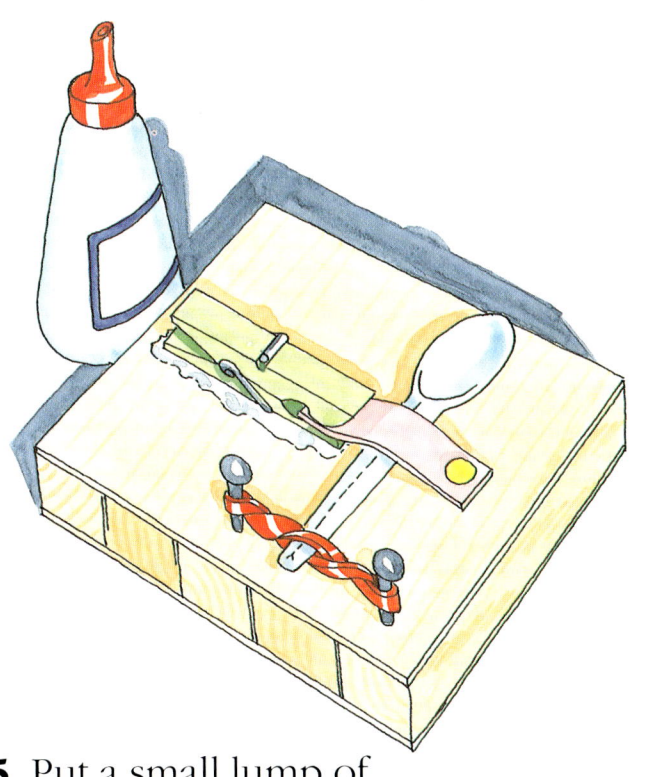

3. Glue the clothespin to the block of wood at a **right angle** to the spoon, as shown. This is the trigger for your catapult.

4. Lay the cardboard strip across the spoon. Use a thumbtack to fasten one end of the strip to the board. Use the clothespin to hold the other end of the strip in place.

5. Put a small lump of modeling clay into the bowl of the spoon. Press the clothespin to fire the catapult. How far does the modeling clay ball go?

6. Experiment with your catapult by winding the rubber band tighter, then looser. Which makes the clay ball go farther? Does using a longer spoon make any difference?

WARNING:
Always be careful when firing your catapult. Never fire heavy or sharp objects. Make sure you will not hit anybody or damage anything.

13

DIGGERS

Diggers and **bulldozers** are used to move dirt and rocks to prepare the ground for new roads and buildings. Some diggers have enormous buckets on them. Have you ever seen a digger at a building site?

A bulldozer can push large amounts of rocks and dirt from place to place.

Making a bulldozer

You will need:

A small shoe box   Paper fasteners
Stiff cardboard   Glue
A ruler   Scissors

1. Cut out a cardboard rectangle 6 inches (15 cm) by 2 inches (5 cm). Fold it lengthwise to make your scoop, as shown.

2. Cut out two cardboard strips 8 inches (20 cm) by 3/4 inch (2 cm). Fold the last 2 inches (5 cm) of each strip, and glue the folded ends to the scoop. Use paper fasteners to attach the other ends of the strips to the box.

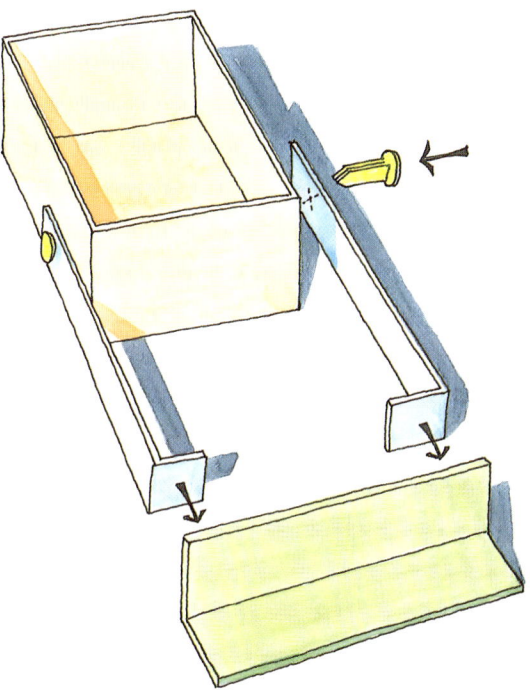

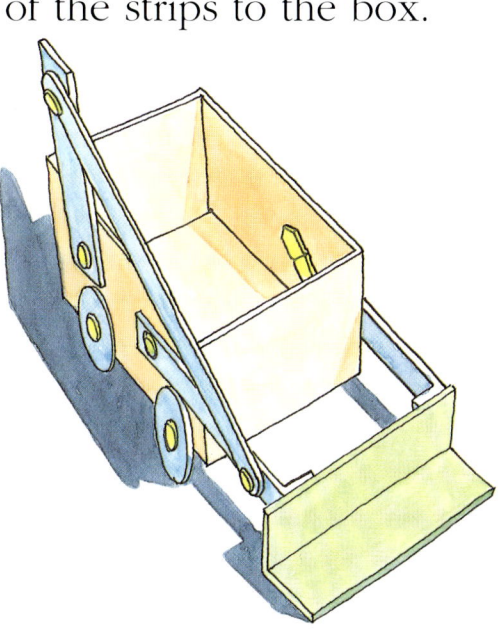

3. Make a handle by cutting a strip of cardboard about 5 inches (12 cm) by 3/4 inch (2 cm). Fasten it to one side of your box, near the back.

4. Measure the distance from the top of the handle to the end of the arm holding the scoop. Cut out a piece of cardboard this length and fasten it at both ends.

5. Cut out four cardboard wheels. Use paper fasteners to attach them to the box.

PULLEYS

Pulleys are used for lifting heavy objects. A pulley is a wheel with a rope around it. By pulling down on one end of the rope, you can lift a heavy load on the other end. Do you think pulling down a heavy object is easier than lifting it up?

Using a pulley

You will need:

An empty matchbox
Three paper clips
Two pieces of nylon
  fishing line
Two rubber bands

Two hooks
A wooden stick
A Styrofoam cup
A clamp attached
  to a stand

Modeling clay
Scissors
A spool
Tape
Water

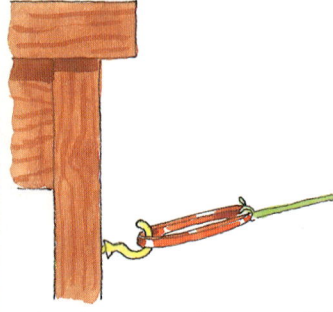

1. Cut a piece of fishing line long enough to fit between two table legs. Tie a rubber band to each end of the fishing line. Tape one hook securely to the leg of each table, as shown. **Do not screw the hooks into the tables**. Attach the rubber bands to the hooks, making sure the line is tight.

16

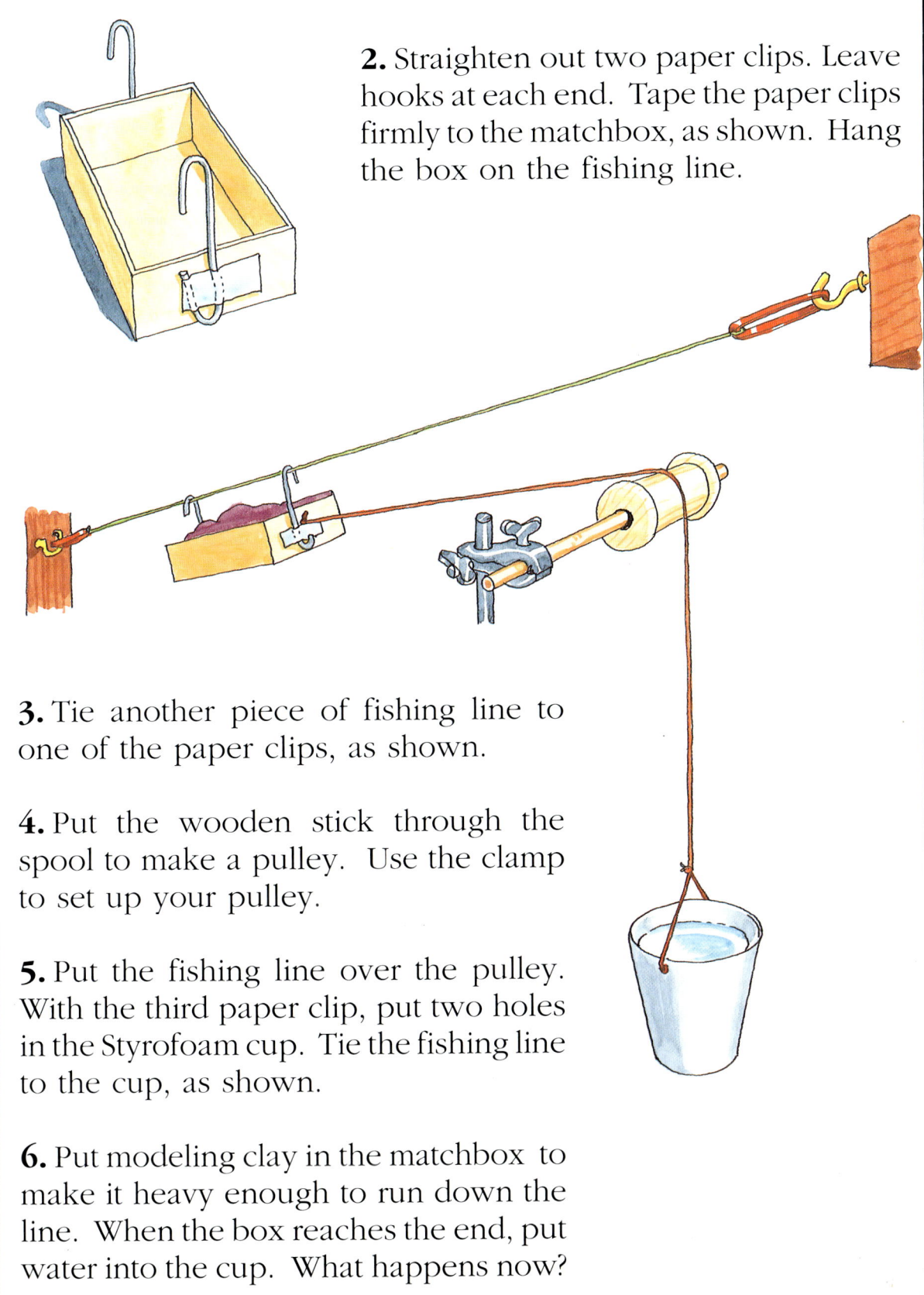

2. Straighten out two paper clips. Leave hooks at each end. Tape the paper clips firmly to the matchbox, as shown. Hang the box on the fishing line.

3. Tie another piece of fishing line to one of the paper clips, as shown.

4. Put the wooden stick through the spool to make a pulley. Use the clamp to set up your pulley.

5. Put the fishing line over the pulley. With the third paper clip, put two holes in the Styrofoam cup. Tie the fishing line to the cup, as shown.

6. Put modeling clay in the matchbox to make it heavy enough to run down the line. When the box reaches the end, put water into the cup. What happens now?

17

CRANES

Cranes use pulleys to lift big, heavy objects. Enormous cranes are used to build tall modern office buildings or load huge containers onto ships.

Cranes are used to help build very tall buildings. Look how tall this crane is.

Making a crane

You will need:

Cardboard   A small cardboard box
Scissors   Glue
Wooden sticks   A hook or magnet
Thread

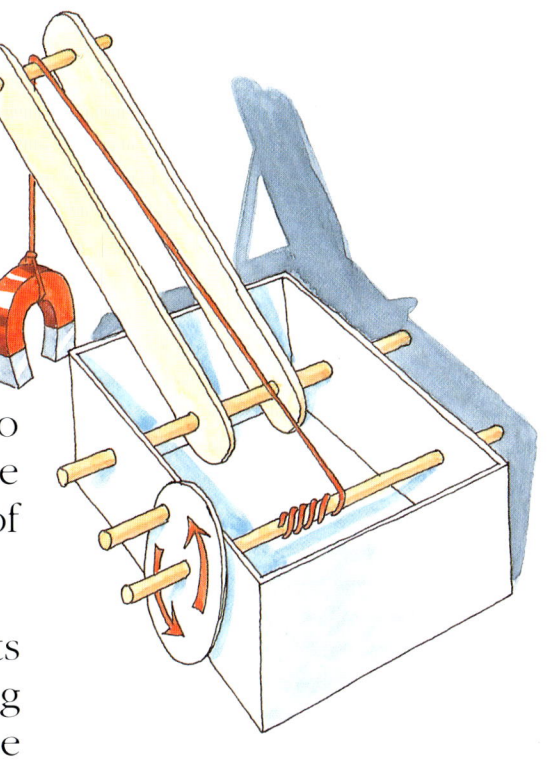

1. Cut two strips of cardboard to make the arm of your crane. Use scissors to make slits in each end of the cardboard strips.

2. Push a short stick through the slits at the end of both strips. Push a long strip through the other slits. Glue the sticks in place.

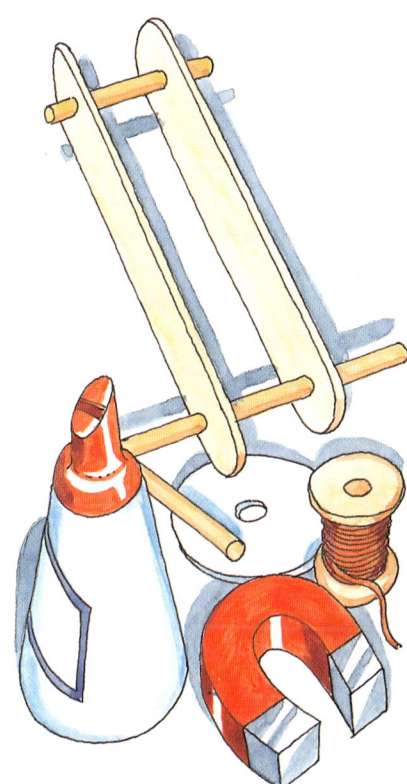

3. Push the ends of the longer stick through the sides of the box, as shown.

4. Push another long stick through the sides of the box, near the back of the crane. Tie some thread to the middle of this stick. Drape this thread as shown.

5. Cut out a small cardboard circle. This is your handle. Attach a short stick near the edge of the circle. Stick the circle onto the end of the long stick at the back of your crane, as shown. Glue it in place.

6. Tie a magnet or hook to the end of the thread. Now try out your crane.

19

BALLOON POWER

Making a dump truck

You will need:

A balloon
Plastic tubing
Tape
A plastic bottle
Scissors
A shoebox with a lid
Two smaller boxes
Cardboard
Four paper fasteners
Glue

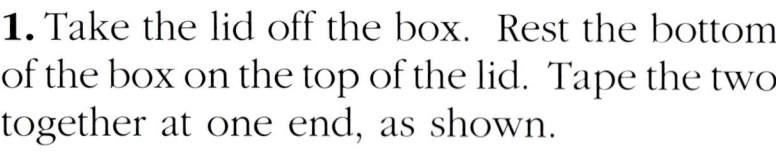

1. Take the lid off the box. Rest the bottom of the box on the top of the lid. Tape the two together at one end, as shown.

2. Stretch the mouth of the balloon tightly around the end of the tubing. Tape it in place so that no air can escape.

3. Glue or tape the smaller boxes to the other end of the lid to make the driver's cab and engine. Cut out cardboard circles for the wheels and attach them to the lid with paper fasteners.

20

4. Make a hole in the box lid. Lay the balloon flat between the box and the lid, as shown. Push the tubing through the hole and out the bottom of the dump truck.

5. Tape the other end of the tubing to the nozzle of the bottle. Squeeze the bottle to blow up the balloon.

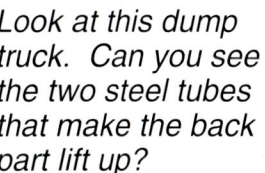

Look at this dump truck. Can you see the two steel tubes that make the back part lift up?

21

DRAGONS

Making a dragon's mouth that snaps

You will need:

A balloon
Plastic tubing
Tape
A plastic bottle
Scissors
A shoebox with a lid
Cardboard
Glue
Egg cartons
String
A paintbox

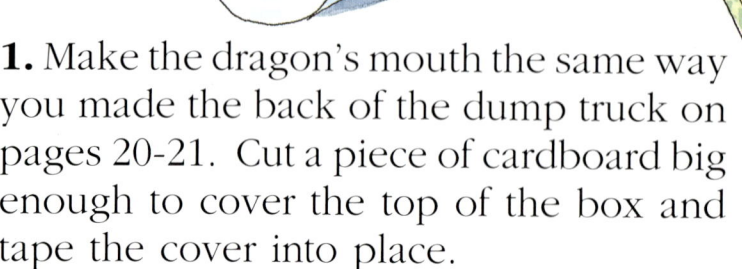

1. Make the dragon's mouth the same way you made the back of the dump truck on pages 20-21. Cut a piece of cardboard big enough to cover the top of the box and tape the cover into place.

2. Cut out cardboard teeth and stick them to the sides of the box. Make the body of the dragon by tying egg cartons together. Decorate your dragon with paint and pieces of cardboard.

3. Squeeze the bottle to open the dragon's jaws. Is your dragon scary?

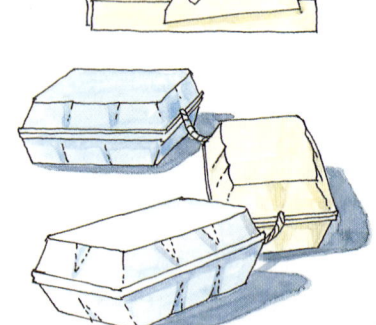

Making the dragon's eyes

You will need:

Two electric bulbs with bulb holders
Insulated copper wires
A small (4.5-volt) **battery**
Tape

WARNING: Power outlets are very dangerous. Never use them in these experiments.

1. Glue or tape the bulb holders where you want the dragon's eyes to be.

2. Tape the battery to the back of the dragon's head. Connect the bulbs and battery with the wire to make an **electric circuit**.

In China and other parts of the world, people hold dragon festivals. They make dragons like this one and dance with them in the street.

23

SWITCHES

Televisions, radios, and electric lights all have switches for turning them on and off. Look around you at home and at school for other machines that have switches. But remember not to play with switches, since electricity can be very dangerous.

There are many switches in an airplane cockpit. See how many you can count in this picture. The pilot must know what each switch is for.

Making a see-saw switch

You will need:

Cardboard
A cardboard box
Paper fasteners
Aluminum foil
A ruler

A 4.5-volt battery
Insulated copper wire
A bulb and bulb holder
Scissors

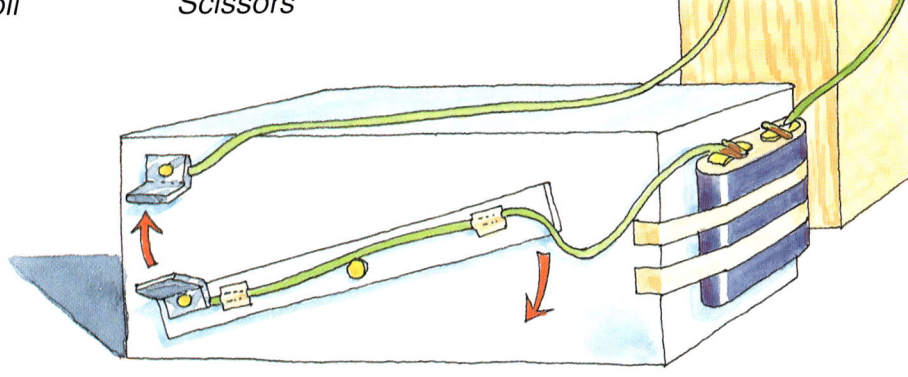

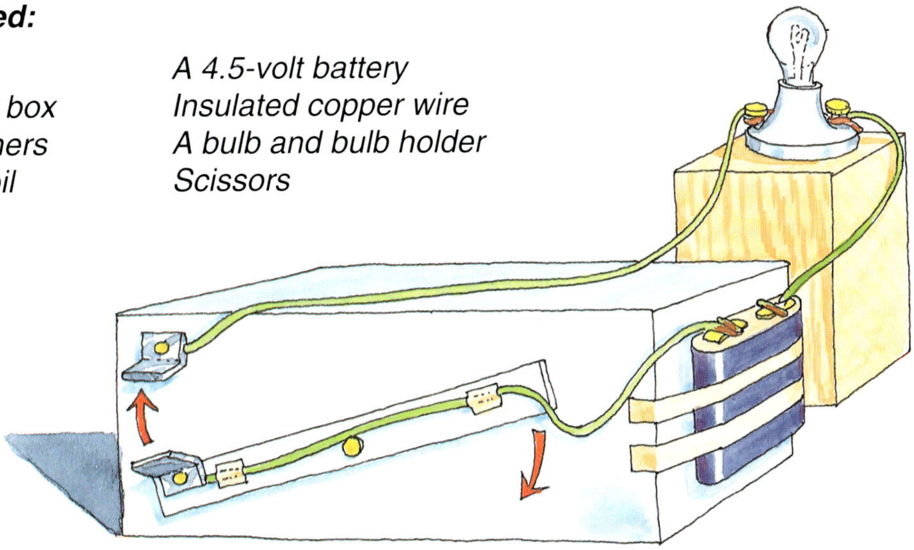

1. Cut a strip of cardboard 5 inches (12 cm) long. Use a paper fastener to attach it to the side of the box, making a kind of see-saw. This is your switch.

2. Fasten a flap of aluminum foil to one end of the switch. Fasten another flap of foil to the box just above it, so that when the switch moves up, the two pieces of foil touch.

3. Wind the end of a piece of wire around each of the fasteners, between the aluminum foil and the cardboard.

4. Tape the battery to one end of the box. Connect the bulb, battery, and see-saw with insulated wire, as shown.

5. Now push the see-saw switch so that the metal flaps touch. The bulb should light up.

WARNING:
Power outlets are very dangerous. Never use them in these experiments.

25

A SWITCH MACHINE

Making an automatic switch machine

You will need:

A large cardboard box
A marble
Cardboard
Thread
A Styrofoam cup

Paper fasteners
A wooden stick
A plastic straw
Tape
Aluminum foil

Insulated wire
A bulb and bulb holder
A 4.5-volt battery
Scissors
Glue

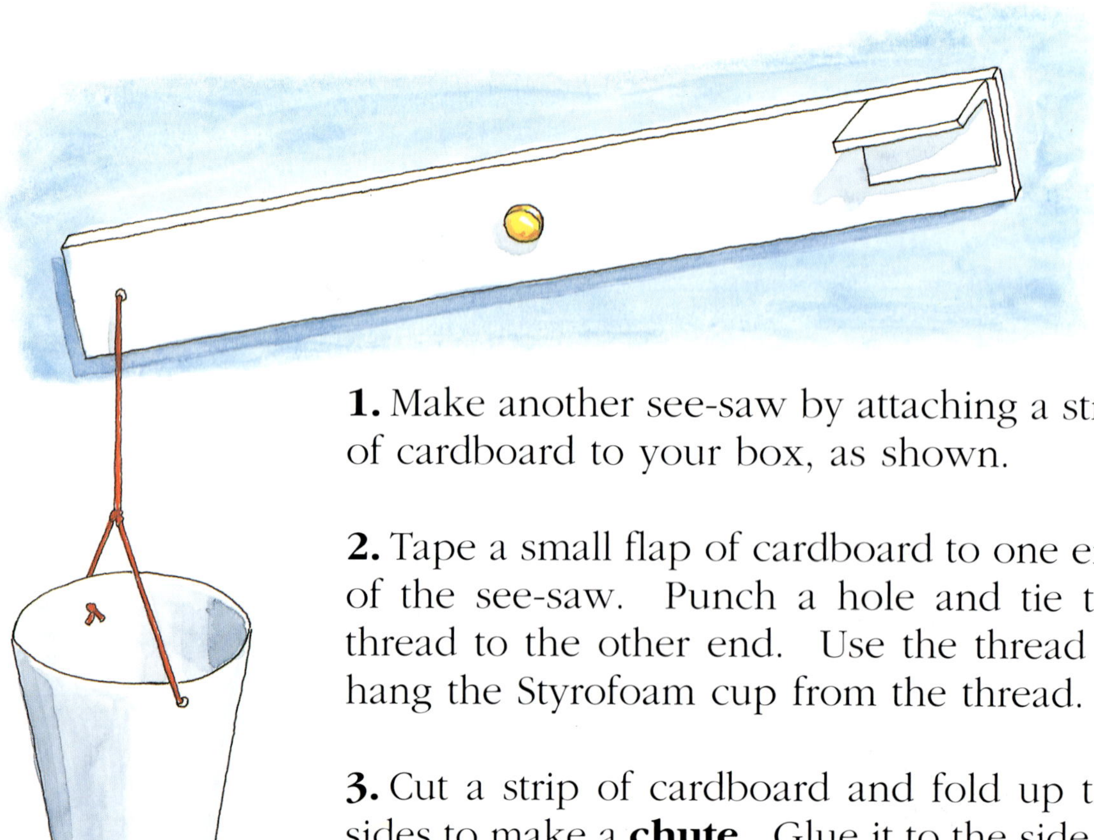

1. Make another see-saw by attaching a strip of cardboard to your box, as shown.

2. Tape a small flap of cardboard to one end of the see-saw. Punch a hole and tie the thread to the other end. Use the thread to hang the Styrofoam cup from the thread.

3. Cut a strip of cardboard and fold up the sides to make a **chute**. Glue it to the side of the box. Place it so the marble will land in the cup when it rolls down the chute.

4. Cut a piece of straw and tape it to the box, directly above the see-saw's flap.

5. Put the wooden stick through the straw so that it rests on the flap. Glue a folded piece of aluminum foil to the other end of the stick. Put the end of a piece of wire between the folds of foil.

6. Fasten a flap of aluminum foil to the box, directly above the top of the stick. Use more wire to connect the two pieces of foil, the bulb, and the battery, as shown.

7. Run a marble down the slope. How many different machines must work before the bulb lights up?

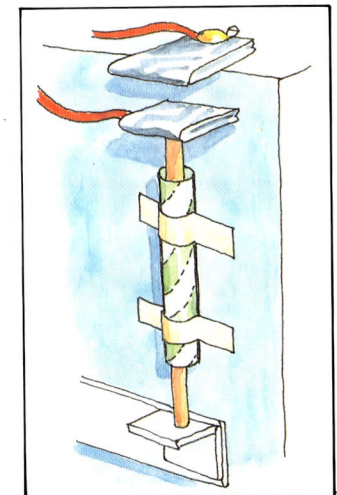

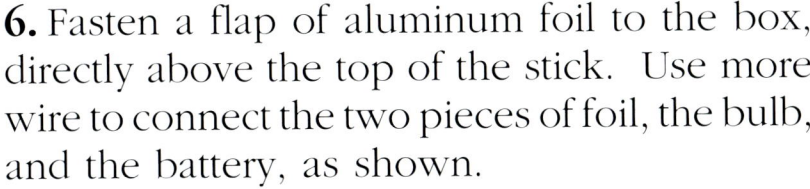

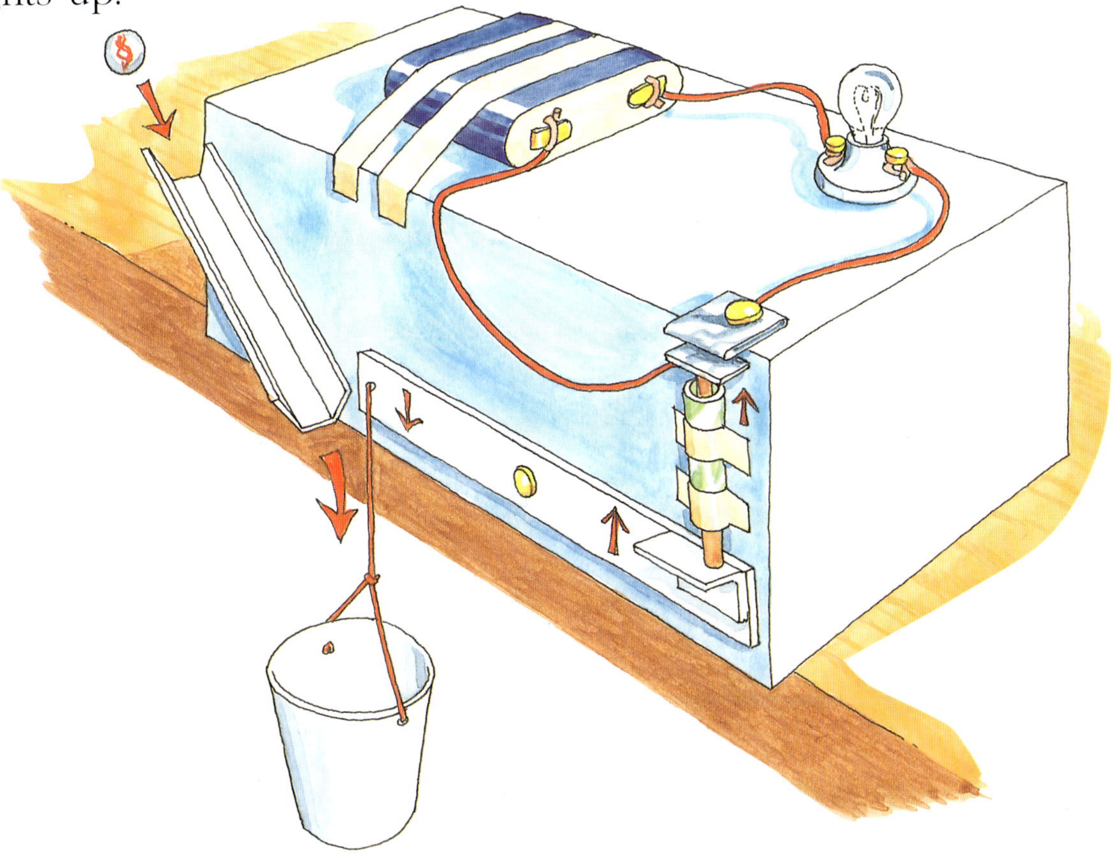

What You'll Need

aluminum foil 25, 26-27
balloon 20-21, 22-23
battery in holder..23, 25, 26-27
block of wood......................12
bulbs in holder ...23, 25, 26-27
cardboard ...5, 7, 9, 10, 12, 15, 19, 20-21, 22-23, 25, 26-27
cardboard boxes 19, 20-21, 25, 26-27
clamp 16-17
clothespin 12
egg cartons 22-23
fishing line, nylon 16-17
glue 10, 12, 15, 19, 20-21, 22-23, 26-27
hammer................................12
hooks 16-17, 19
magnet 19
marble 26-27
matchbox, empty 16-17
modeling clay 12, 16-17
nails 12
paintbox 5, 7, 22-23
paper clips......................16-17

paper fasteners 5, 7, 9, 10, 15, 20-21, 25, 26-27
pencil 5, 7, 9, 10
plastic bottle 20-21, 22-23
plastic spoon 12
plastic straw 26-27
plastic tubing 20-21, 22-23
rubber bands 9, 12, 16-17
ruler 5, 7, 9, 10, 12, 15, 25
scissors...5, 7, 9, 10, 15, 16-17, 19, 20-21, 22-23, 25, 26-27
shoebox 15, 20-21, 22-23
spool 16-17
stand 16-17
string 22-23
Styrofoam cup......16-17, 26-27
tables 16-17
tape 9, 16-17, 20-21, 22-23, 26-27
thread 19, 26-27
thumbtack............................12
wire, insulated 23, 25, 26-27
wooden sticks 16-17, 19, 26-27

More Books About Machines

Baxter's Book of Machines. Leon Baxter (Ideals)
The Book of Foolish Machinery. Donna L. Pape (Scholastic)
Diggers and Loaders. Graham Thompson (Gareth Stevens)
In the Factory. Malcolm Dixon (Franklin Watts)
Machines: Detecting Sequence. Johanna P. Pomeroy (Educational Activities)
Machines As Tall As Giants. Paul Strickland (Random House)
Monster Machines. Paul Nash (Garrett Educational)
See How It Works: Earth Movers. Tony Potter (Macmillan)
Simple Machines. Anne Horvatic (E. P. Dutton)
The World of Machines. (Raintree)

More Books With Projects

Boden's Beasts. Arthur Boden and John Woodside (Astor-Honor)
Experimenting with Batteries, Bulbs, and Wires. Alan Ward (David & Charles)
The KnowHow Book of Batteries and Magnets: Safe and Simple Experiments, Models, and Games. H. Amery and A. Littler (EDC Publishing)
The KnowHow Book of Paper Fun: Lots of Things to Make from Paper and Card. A. Curtis and J. Hindley (EDC Publishing)
Make It with Odds and Ends! Felicia Law (Gareth Stevens)
Papercrafts. Judith H. Corwin (Franklin Watts)
Switch On, Switch Off. Melvin Berger (Harper & Row)
Things I Can Make with Paper. Sabine Lohf (Chronicle Books)

Places to Write for Science Supply Catalogs

Suitcase Science
Small World Toys
P. O. Box 5291
Beverly Hills, California 90209

The Nature of Things
275 West Wisconsin Avenue
Milwaukee, Wisconsin 53203

Adventures in Science
Educational Insights
19560 Rancho Way
Dominguez Hills, California 90220-6038

Sargent-Welch Scientific Company
7300 North Linden Avenue
Skokie, Illinois 60076

GLOSSARY

battery
A device that makes or stores electricity.

bulldozers
Powerful tractors with blades on the front, used for moving soil and rocks.

chute
A narrow, gutter-shaped slide.

diggers
Heavy machines used to scoop out holes.

electric circuit
A loop of wires and objects connected so that electricity will flow through it.

factories
Buildings containing complicated machines that are used for making things.

fulcrum
The point on which a lever balances or turns.

joint
A place where two or more things, like pipes, come together.

lever
A bar used for moving heavy weights. You push or pull on one end of the bar to lift the weight on the other end.

pulley
A wheel with a rope around it. Pulleys make heavy objects easier to lift.

right angle
A square corner, like the corner of a book or a box, where two lines meet.

Picture acknowledgements
The publishers would like to thank the following for allowing their photographs to be reproduced in this book: Cephas Picture Library, p. 24 (Nigel Blythe); Eye Ubiquitous, p. 6 (J. Winkles); Hutchison Library, p. 23 (Leslie Woodhead); Sefton Photo Library, pp. 11, 18; Topham Picture Library, pp. 8, 21; Zefa, p. 4 (T. Horowitz), p. 14 (J. Pfaff). Cover photography by Zul Mukhida.

INDEX

airplane cockpit 24
automatic switch machine 26-27

balloons 20-21, 22
batteries 23, 25, 26-27
bodies 8
bones 8
building sites 11, 14, 18
bulldozers 14-15

cat, model 5
catapults 12-13
China 23
computers 4, 6
cranes 18-19

diggers 14
dragon, model 22-23
dragon festivals 23
dump truck 20-21, 22

electric circuit 23
extenders 10-11

factories 4
fulcrums 8

joints 8

legs 5, 8-9
levers 4, 6, 8, 11
lights, electric 24

model trains 6

pulleys 4, 16-17, 18

radio 24
railroad signals 6-7
right angle 13
robots 8

see-saw switches 25, 26-27
switches 24-27

televisions 24

FITCH STREET SCHOOL LIBRARY